# Keyboard Success

## Student Flip Book

## Second Edition

**Sam Miller • Mary Smith • Ann Fidanque • Gail Sullivan**

© International Society for Technology in Education, Second Edition, 2000

## Keyboard Success Student Flip Book, Second Edition

**Director of Publishing**
Jean Marie Hall

**Project Coordinator**
Diannah Anavir

**Acquisitions Editor**
Anita McNear

**Copy Editors**
Ron Renchler, The Electronic Page
Maggie Wheeler

**Technical Editor**
Corinne Tan, Tracy Cozzens

**Cover & Book Design**
Sue Roberts

**Hand Model**
Kye Thomas Ruddell, age 10

Copyright © 2003 International Society for Technology in Education

World rights reserved. No part of this book may be reproduced or transmitted in any form or by any means—electronic, mechanical, photocopying, recording, or otherwise—without prior written permission of the publisher. For permission, ISTE members contact Permissions Editor, ISTE, 480 Charnelton Street, Eugene, OR 97401-2626; fax: 1.541.434.8943; e-mail: permissions@iste.org. Nonmembers contact Copyright Clearance Center, 222 Rosewood Drive, Danvers, MA 01923; fax: 1.978.750.4744.

An exception to the above statement is made for K–12 classroom materials or teacher training materials contained in this publication and, if applicable, on the accompanying CD-ROM. A single classroom teacher or teacher educator may reproduce these materials for his or her classroom or students' uses. The reproduction of any materials in this publication or on the accompanying CD-ROM for an entire school or school system, or for other than nonprofit educational purposes, is strictly prohibited.

*Trademarks:* Rather than put a trademark symbol in every occurrence of a trademarked name, we state that we are using the names only in an editorial fashion and to the benefit of the trademark owner, with no intention of infringement of the trademark.

**International Society for Technology in Education (ISTE)**
480 Charnelton Street
Eugene, OR 97401-2626
Order Desk: 1.800.336.5191
Order Fax: 1.541.302.3778
Customer Service: orders@iste.org
Books and Courseware: books@iste.org
Permissions: permissions@iste.org
World Wide Web: www.iste.org

# Second Edition
ISBN 1-56484-153-7

# Table of Contents

**Welcome to Keyboard Success!** ............ iv

Lesson 1 ............ 1
Lesson 2 ............ 2
Lesson 3 ............ 3
Lesson 4 ............ 4
Lesson 5 ............ 5
Lesson 6 ............ 6
Bonus Lesson 6 ............ 7

Lesson 12 ............ 18
Lesson 13 ............ 20
Lesson 14 ............ 22
Lesson 15 ............ 24
Lesson 16 ............ 26
Bonus Lesson 12 ............ 19
Bonus Lesson 13 ............ 21
Bonus Lesson 14 ............ 23
Bonus Lesson 15 ............ 25
Bonus Lesson 16 Keyboarding Rate ............ 27

Lesson 22 ............ 39
Lesson 23 ............ 41
Lesson 24 ............ 43
Lesson 25 ............ 45
Lesson 26 ............ 47
Bonus Lesson 22 ............ 40
Bonus Lesson 23 Keyboarding Rate ............ 42
Bonus Lesson 24 ............ 44
Bonus Lesson 25 ............ 46
Bonus Lesson 26 ............ 48

**Lesson 7** ............................... **8**

   Bonus Lesson 7
   Keyboarding Rate ................. 9

**Lesson 8** ............................. **10**

   Bonus Lesson 8 .................... 11

**Lesson 9** ............................. **12**

   Bonus Lesson 9 .................... 13

**Lesson 10** ........................... **14**

   Bonus Lesson 10 .................. 15

**Lesson 11** ........................... **16**

   Bonus Lesson 11 .................. 17

**Lesson 17** ........................... **29**

   Bonus Lesson 17 .................. 30

**Lesson 18** ........................... **31**

   Bonus Lesson 18 .................. 32

**Lesson 19** ........................... **33**

   Bonus Lesson 19 .................. 34

**Lesson 20** ........................... **35**

   Bonus Lesson 20 .................. 36

**Lesson 21** ........................... **37**

   Bonus Lesson 21 .................. 38

**Lesson 27** ........................... **49**

   Bonus Lesson 27 .................. 50

**Lesson 28** ........................... **51**

   Bonus Lesson 28 .................. 52

**Lesson 29** ........................... **53**

   Bonus Lesson 29 .................. 54

**Lesson 30** ........................... **55**

   Bonus Lesson 30
   Keyboarding Rate ............... 56

# Welcome to Keyboard Success!

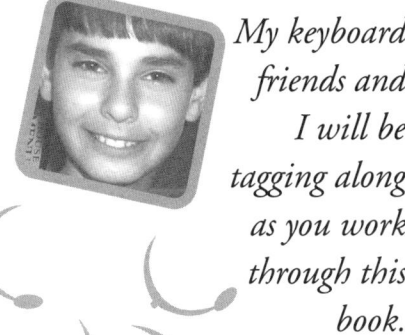

*My keyboard friends and I will be tagging along as you work through this book.*

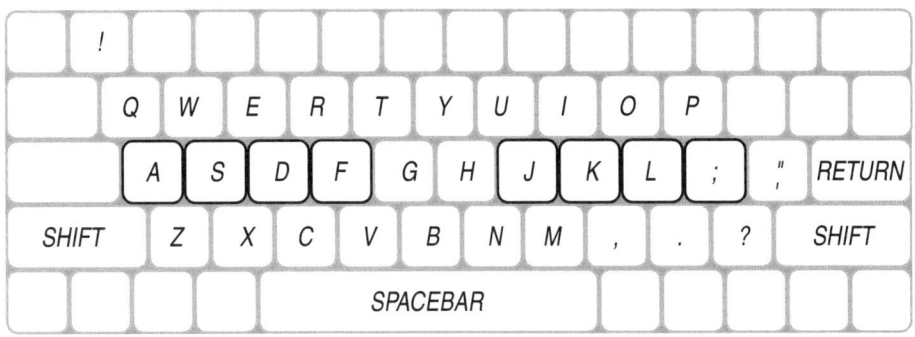

There's a lot to learn! Like how to sit at a computer or typewriter correctly and how to position your hands on a keyboard. You'll find out about home keys and which fingers to use for which keys. Soon you'll be keyboarding. What an adventure! You're on your way to …

**Keyboard Success!**

# The End!

## Happy Keyboarding!

# Lesson 1 — Position Chart for Good Keyboarding Habits

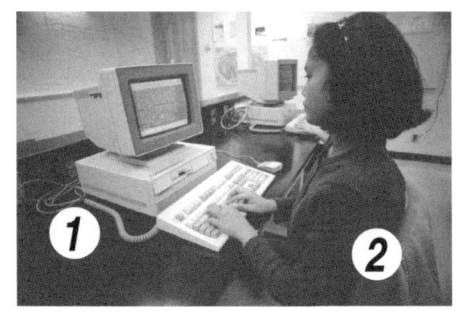
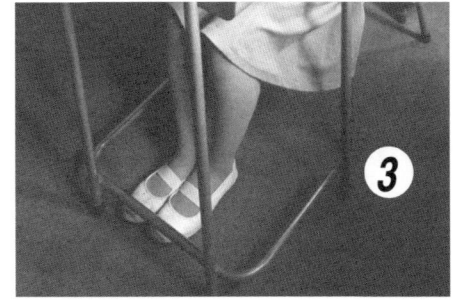

**1 CLEAR**
work area

**2 SEAT**
- sit up straight
- elbows in
- one handspan away

**3 FEET**
and chair
flat on floor

**4 FINGERS**
- wrists flat
- fingers curved
- tips on keys

# Bonus Lesson 30 • Keyboarding Rate

Did the quick, brown fox jump?

Is there time for some people?

Magician Vic can zap all yaks.

# Lesson 2

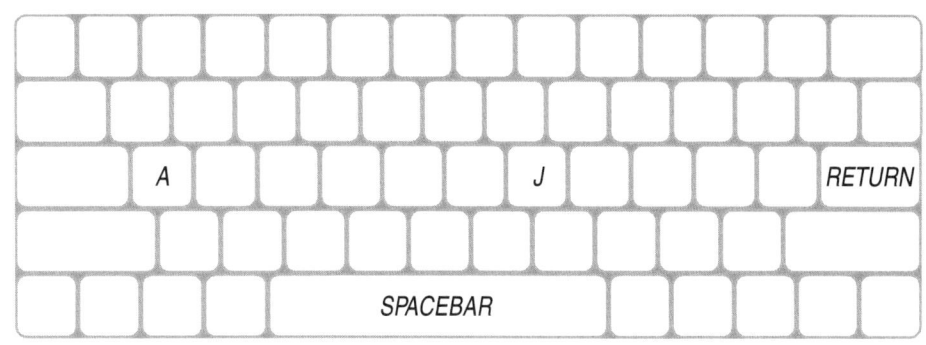

aaa jjj aaa jjj aaa jjj aj aj ja ja a ja

ja ja ja aj aj aj a ja a ja jaa jaa jaj jaj

jaa jaa jaa ajj ajj ajj jaj jaj jaj aja aja

# Lesson 30

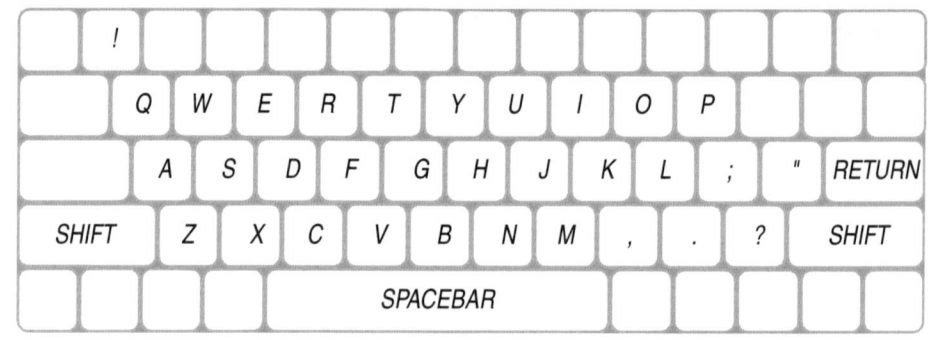

Big planes can't zoom by the new tower.

Didn't my ax zip through the black wood?

Won't Elizabeth, John, and Mark report?

# Lesson 3

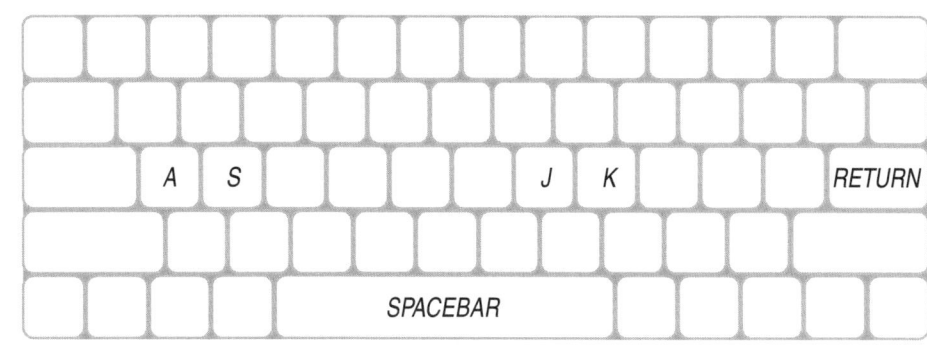

jak a jaks sak sks aks aka ksk jka jaj

sass jak ask a sak as a ask a as a ask a

kass asks ask kass jask sass kass jass

# Bonus Lesson 29

Vivian's pet viper wouldn't come home. "This isn't like my viper. He avoids leaving his cave," she said. Victor's brother said, "Maybe your viper's the vampire's next victim."

# Lesson 4

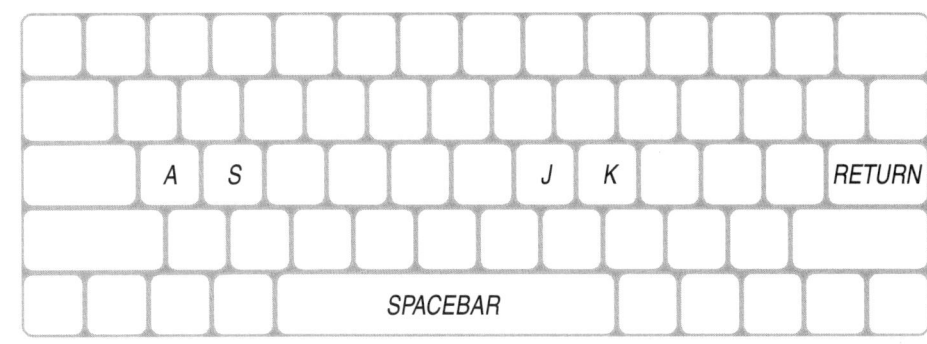

ask a as a ask a as a sak a sak kak

sass kas sass jass kass kjk ksk kak

ak ak jak jak sak sak ask ak ask kass

# Lesson 29

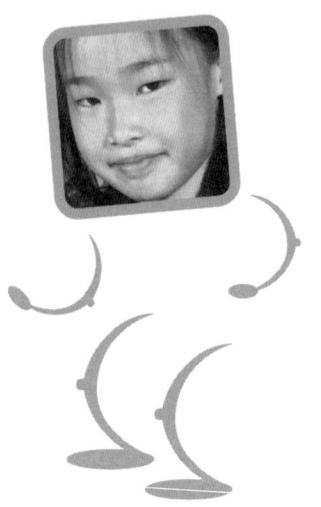

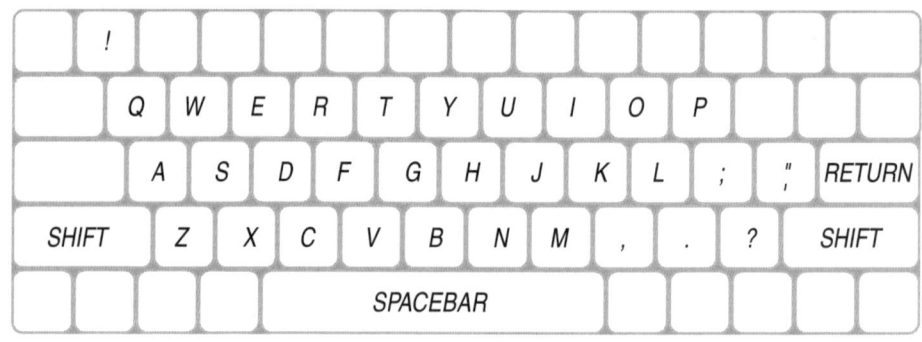

**Vicky's friend Steve will vote for Val.**

**Isn't Vic's brother driving? Let's go.**

**Val's van has five shelves, doesn't it?**

# Lesson 5

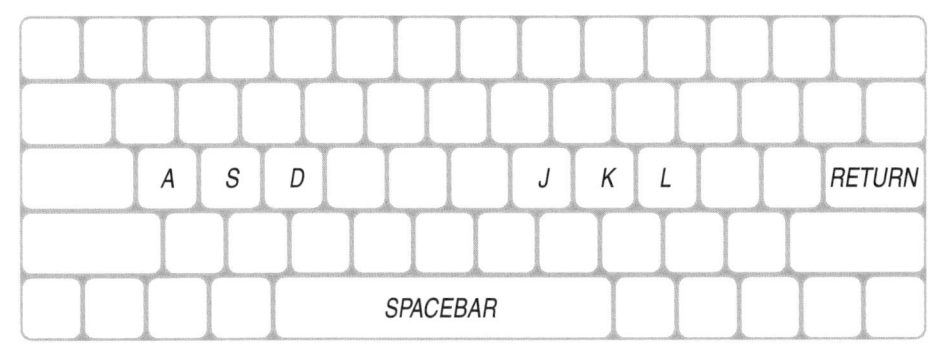

lass adds a salsa salad a sad dad

alas a dallas a sad kass ads a lad adds

ask lads ask a lass all dads add salsa

# Bonus Lesson 28

Mazie spotted a squirrel squirming in a trap. She quickly freed the squirrel. Quietly, she spoke to it, "You will be quite well in no time."

# Lesson 6

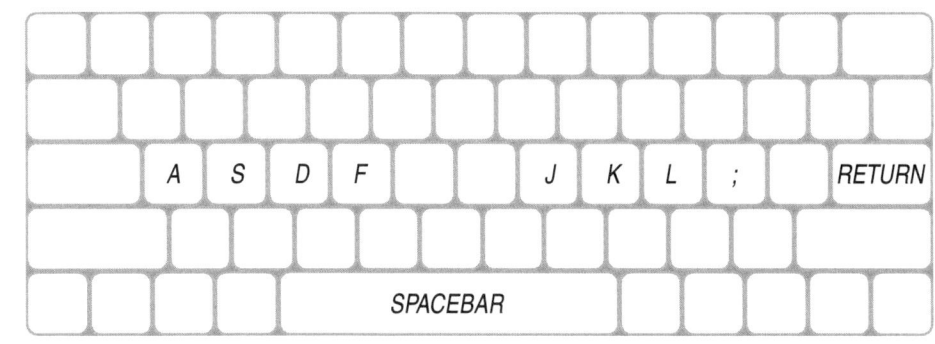

ask dad; ask a lass; dads salads; ads;

a fall fad; a flask falls; fall flasks;

ask all lads; dads fads; salad fads;

# Lesson 28

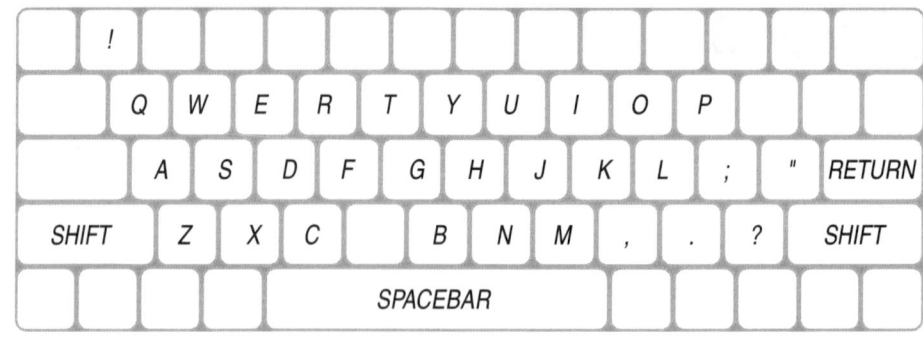

**Queen Janice quarreled about the quarter.**

**Gwen yelled, "Quit those quiz questions."**

**"A square of my quilt qualified to win."**

# Bonus Lesson 6

dads flask; as a lad; sass a lass; salal;

salal salads; all fall; a fall ad; flasks fall;

a sad lad falls; a lads salsa salad;

# Bonus Lesson 27

Zola plans to run away to a city in Tanzania. She will carry her zither in a yellow suitcase. She hopes to get a job in that country. We wish her lots of luck.

# Lesson 7

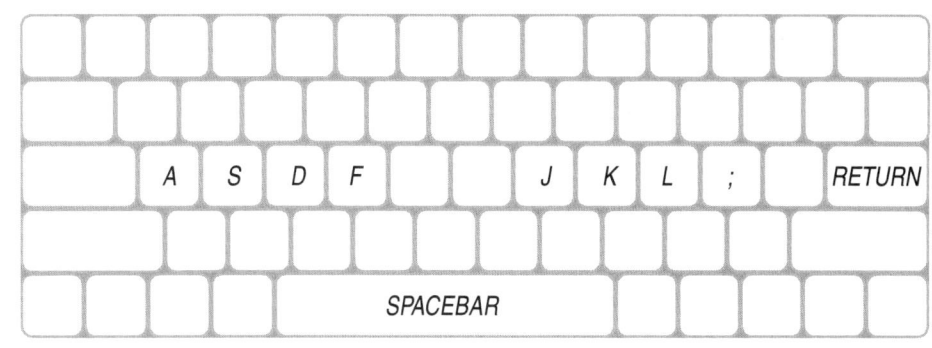

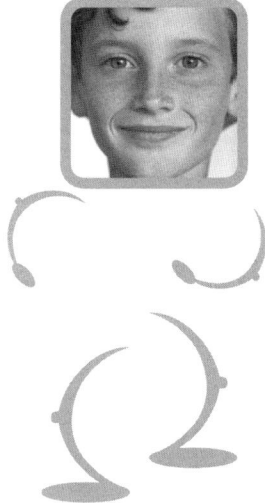

ask dad; ask a lass; a lads salal salad

fall flasks; fall fads fall; a fall fad;

a sad lass falls; alas a lad; salad fads;

# Lesson 27

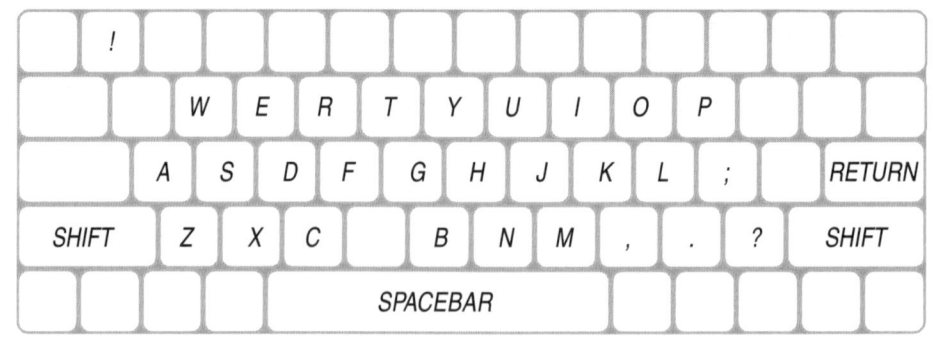

Elizabeth heard jazz while eating pizza.

Do you play in yucky, messy alleys?

The lazy, dozing lizard amazed Larry.

# Bonus Lesson 7 • Keyboarding Rate

all lads salsa salsa; ask lass

flask falls; sad jak; all alas

dad adds salsa; lads alas dads

# Bonus Lesson 26

The word dinosaur means terrible lizard. Pterodactyls were flying dinosaurs. Many people do not realize that they did not really fly, they glided. The most famous and amazing dinosaur was Tyrannosaurus Rex.

# Lesson 8

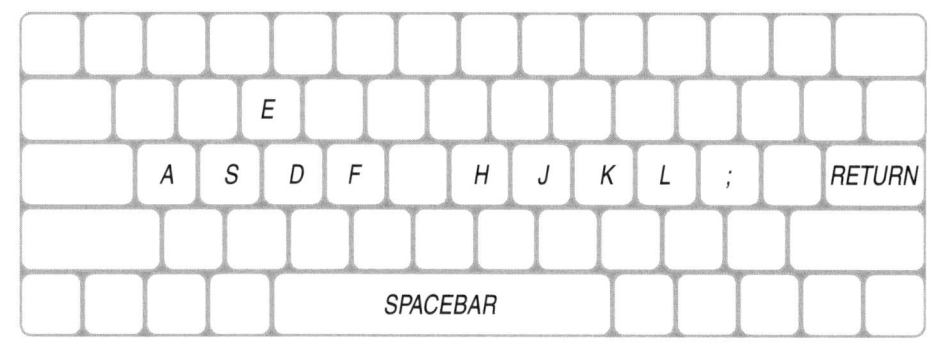

dad led elks; feed eels leeks; seals fade;

he had; held half; has a heel; she heals;

seal a flask; as she ladles; dead ahead;

# Lesson 26

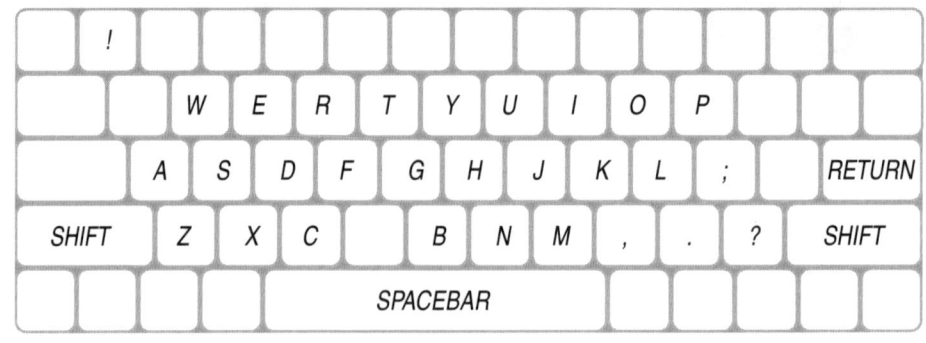

Yes, you ran all the way to Yellowstone.

Liz did crazy, zany zigzags with pizzazz.

Way yonder a lazy zombie yodeled loudly.

# Bonus Lesson 8

false shells; led a sale; half a salad

eels held shells; jell a salad; a fake leaf

has jade sales; a shell shed; a leaf fell

# Bonus Lesson 25

Patrick picked a piece of pineapple upside down pie. His pal Alexis was keeping the extra pan of pie for a surprise. Alexis said to Patrick that he should keep his paws out of her pie pan.

# Lesson 9

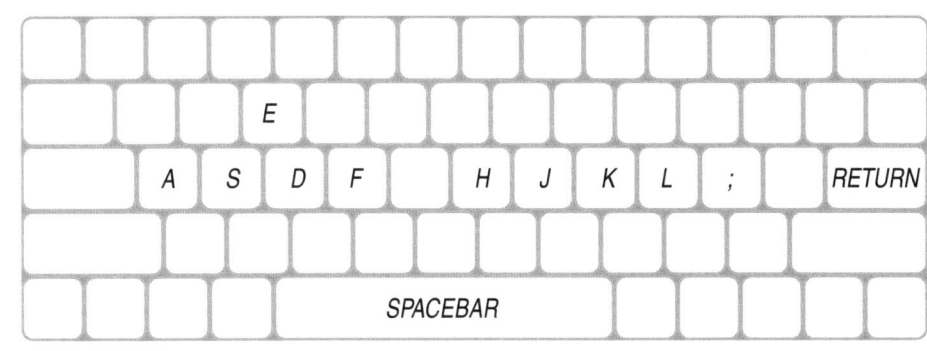

elks seek; seals feed; a fake dead eel;

ha ha hee hee; seek shade; sea shells;

she had asked; safe lakes; a half fed elf;

# Lesson 25

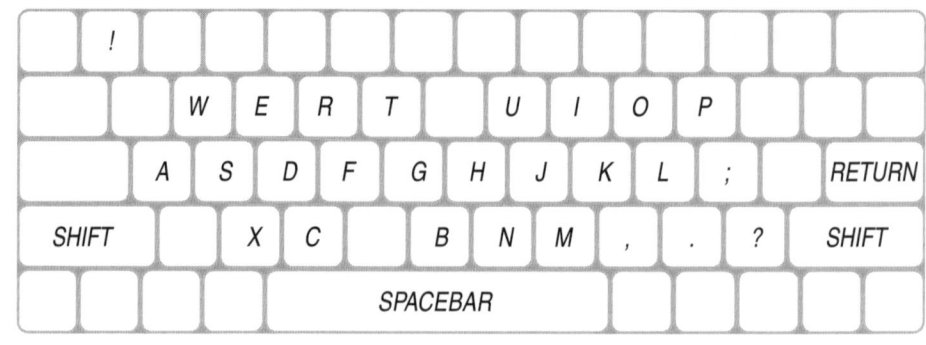

People plant peach, pear, and plum trees.

Max plans to relax in Mexico or Texas.

Will Patti take a maximum of six exams?

# Bonus Lesson 9

lake shade; false sashes; a salad ad

she heals elks; a desk safe; heed a deed

see sea shells; a deaf elf; has a hall

# Bonus Lesson 24

Pam planned a perfect trip to Eexnon planet. On the road Max shouted for her to stop and help pull the trapped taxi out. Then Pam and Max relaxed with space popsicles.

# Lesson 10

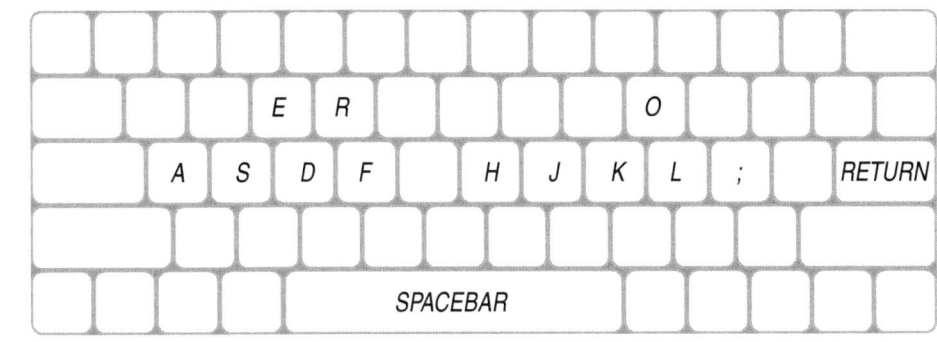

old fool; hello floor; look offshore;

red rose; share her doll; real red rakes;

look for food; reorder jars; old hoses;

# Lesson 24

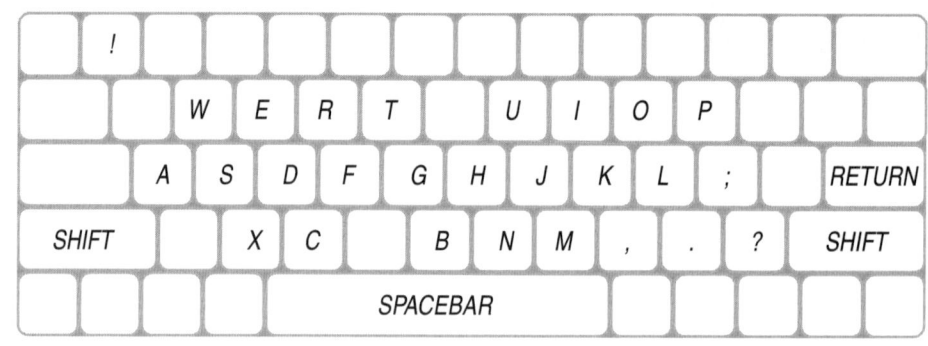

Max and Tex mix sixteen boxes of jello.

Rex brought a purple puppet from Mexico.

Pearl, will Pam please pass the peas?

## Bonus Lesson 10

her old oak desk; hear a hare; ask dad;

soak her rash; a hard floor; off of a roof;

for a shelf; rah rah rah; rake so hard;

# Bonus Lesson 23 • Keyboarding Rate

Her cat Jack sits at the door.

Kurt rushed to see the circus.

Lee skis faster to catch Karl.

# Lesson 11

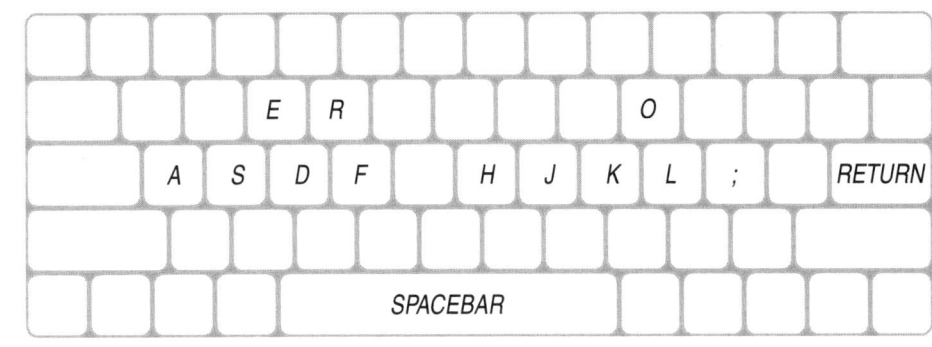

old jello; fool a horse; old hoses; hello;

real red roses; ears hear larks; sold food;

look for rolls; dark lakes; order shoes;

# Lesson 23

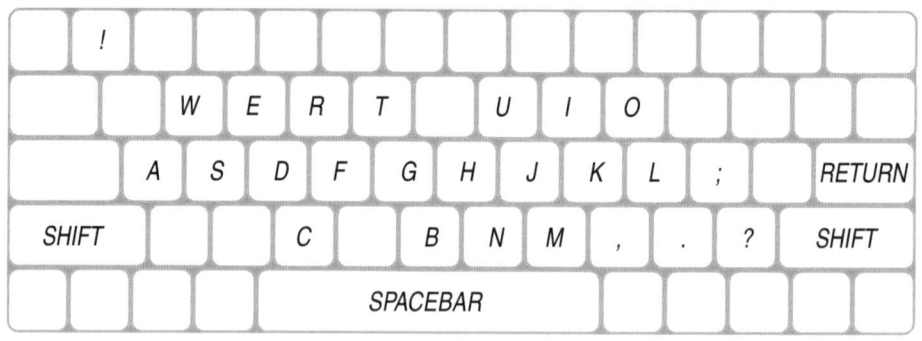

Mama and Marie made a major mistake!

Babies babble; black bumble bees bite.

Am I unable to remember his best fables?

# Bonus Lesson 11

read for her; hear jokes; sharks food;

sold jello; a door hole; a red jar;

look for desks; red shells; old ladders;

# Bonus Lesson 22

Ben is running in the Butte to Butte race now. He scrambled to the mob in the middle. Bill had been in back of Ben but Bill broke his Butte to Butte record.

# Lesson 12

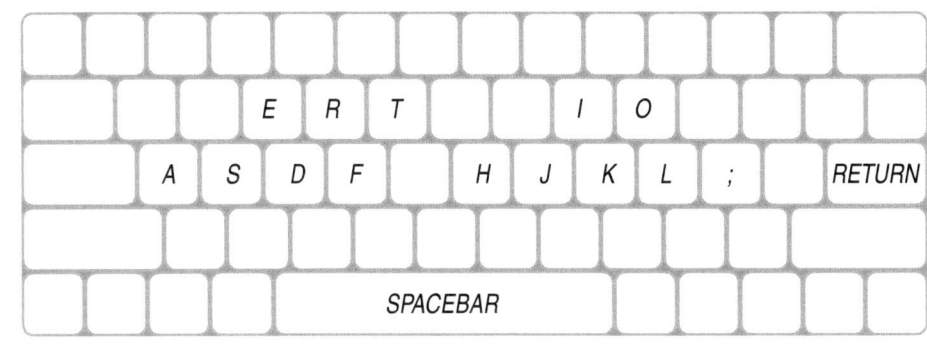

if it is; like to ride; the kite flies;

a toad tried; ate it at first; hire kids;

take to the trail; start to slide the stilts;

# Lesson 22

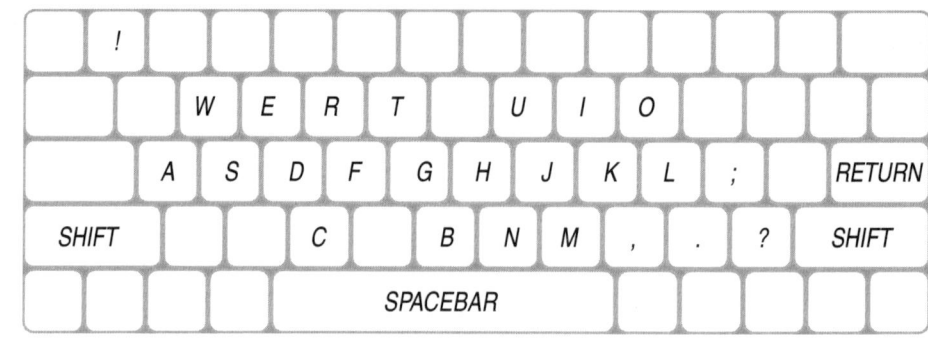

**Mama melted marshmallows. Make nine!**

**Bob Budd hobbled to the Scrabble board.**

**Mom found brontosaurus fossils in Maine.**

# Bonus Lesson 12

three trolls; a dotted toad; too hot

tried to slide; fifth sister; it fits her;

hoist sails; stare at stale fried fish

# Bonus Lesson 21

Alan guarded the fort in the jungle forest. He heard a long, loud growl, the sound of the hungriest tiger. Alan signaled Tess to close the gate.

# Lesson 13

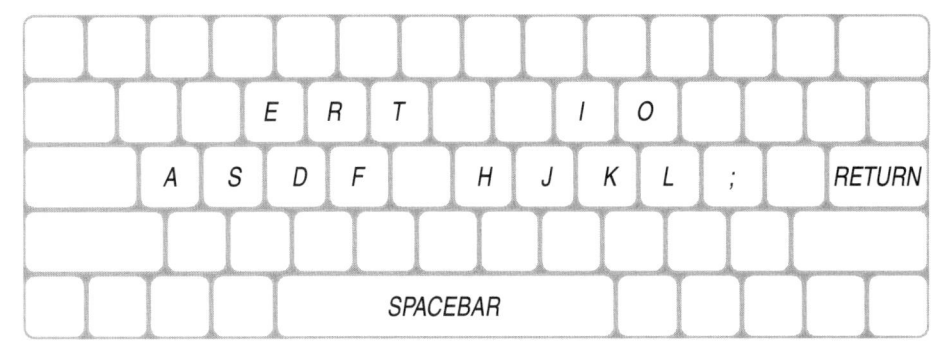

if it sits; this state; lots of fir trees;

their last letter; at the third hit;

tried the third trail; ride the first jet;

# Lesson 21

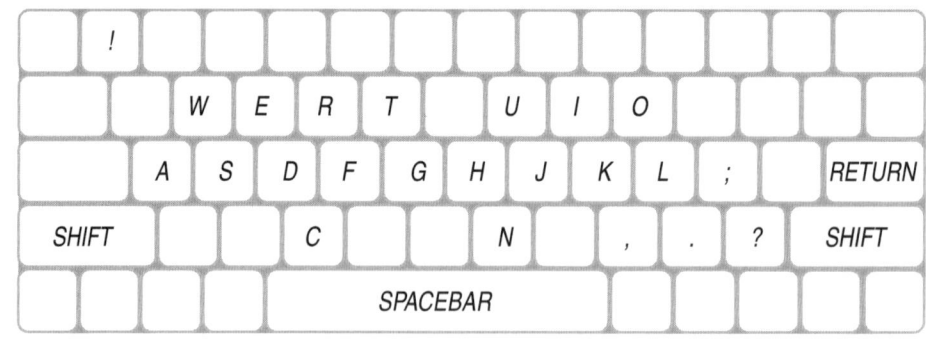

Roger is not going. Ted gets to go now!

Where is Sarah? She giggles with Gina.

Don gets a goat. Sue juggles oranges.

# Bonus Lesson 13

slide it to the lake; at the oak stairs

is sort of dried; tore his toe; lost tot

trotted to the hill; this is; she said

# Bonus Lesson 20

Gerald and his gang created a great show in his garage. Together the gang thought of the title, Dracula and King Kong Go to the Cascades.

# Lesson 14

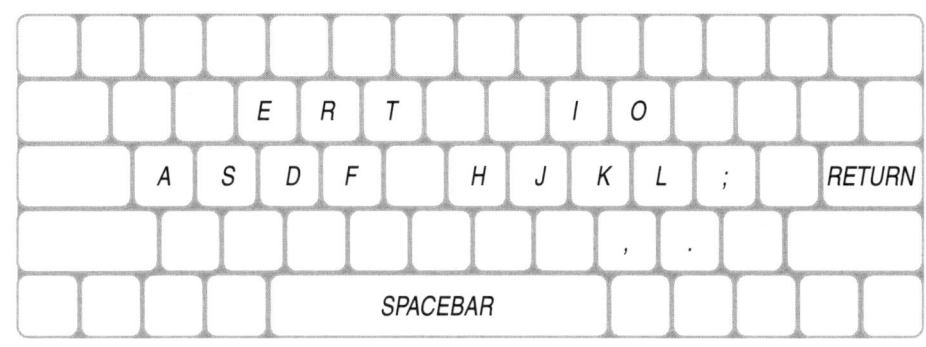

the hot jail. the other three. if he does.

a, s, d, f, j, k, l. a, s, d, f, j, k, l. letters

had a flat tire. tore his toe, lie still.

# Lesson 20

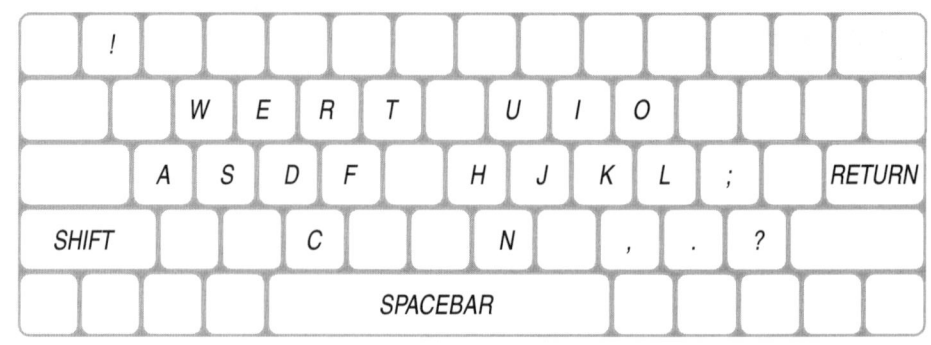

Gigi gags the goose. The goslings go!

When will Elsa grin? Gina is digging.

Tigger is a tiger. Greta wears goggles!

# Bonus Lesson 14

three ft. the hr. karate. hotels.

trolls, trails, trees, dads, lads, lasses

three little fish. likes to slide those stilts.

# Bonus Lesson 19

I like hot weather. Ike likes to drink cola and eat snickerdoodles when it is hot. Jane likes to drink ice water. Neena likes to read and write stories. Nadine earns nickels.

# Lesson 15

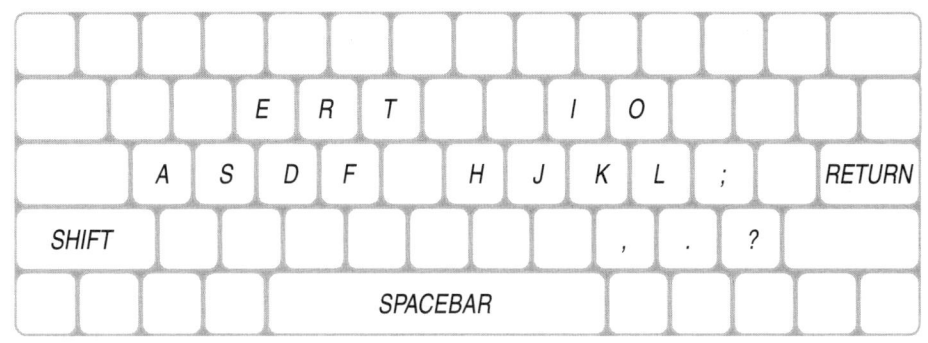

Oskar ate his toast. Is Kerri tired?

Is that a toad? I still see Oak Lake.

Had Lisa, Kae, or Joel lost? Is it his?

# Lesson 19

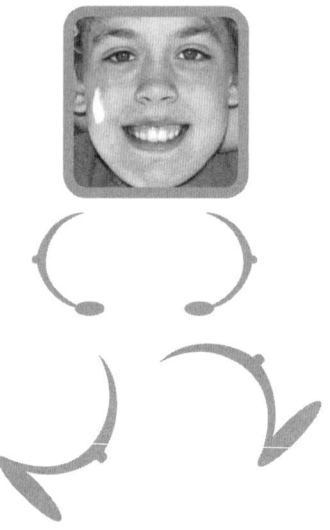

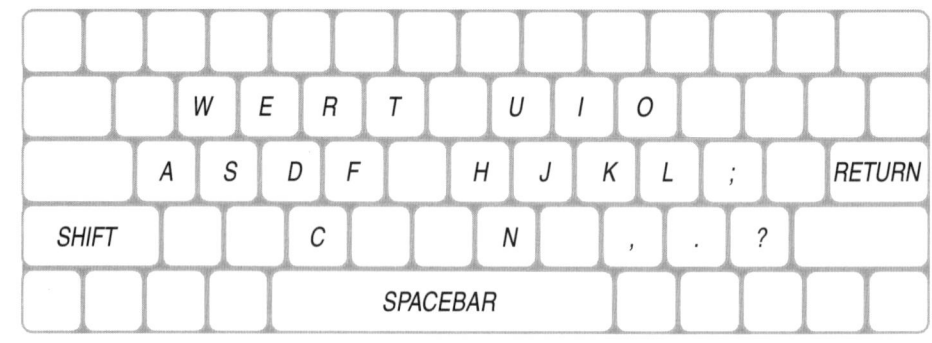

Nonie cannot land one. Ned wants Nikes.

He will know the weather. How did I win?

Our cow chewed her cud. Nan is a friend.

# Bonus Lesson 15

Lois skied to Hero Lake at three to see Kerri. Herds of Elk trotted to Hero Lake too. Lisa stared at the herds as Kerri skied the other trail.

# Bonus Lesson 18

Here are our stock car races. Jarl Jute, a cool dude, has oil or fuel for the autos. Here are four fast cars. Her car is fifth.

# Lesson 16

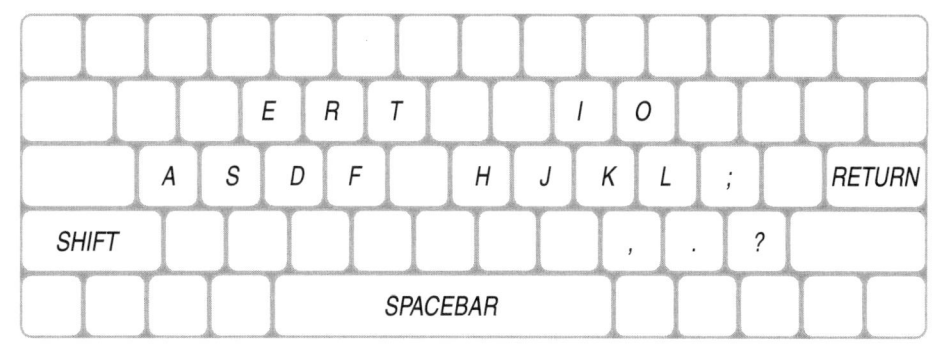

Is Jed late for his date, Karla Oakes?

Look after Kelli Kale. Ollie Hill hides.

O.K., Lissa, or Jake hate hot fried eel.

# Lesson 18

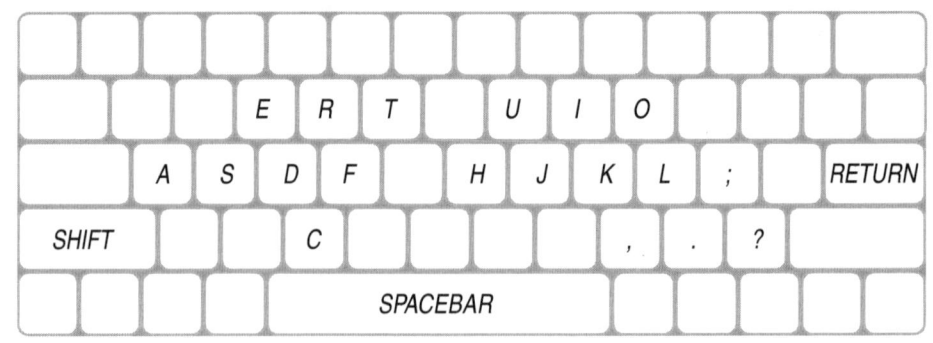

I hear our ukulele call. I trust Lulu.

Is the circus here? Ice Lake is so cold.

Jack tells such useless facts. It is ice.

# Bonus Lesson 16 • Keyboarding Rate

her hero dreads; so read jokes

flasks of roses; rare jar fell

hold her ladder; a lark soars

# Bonus Lesson 17

Jud sees a letter of clues at the cafe. He tried to crack the code. It said to rush to the roller coaster at the surf.

**Time to Flip the book!**

**Flip this page, then turn the whole book around.**

# Lesson 17

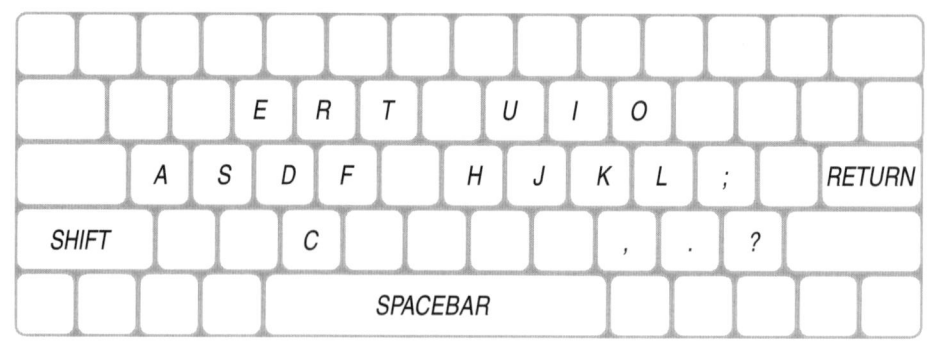

Kris sees crackers at the couch.

Judd uses flour to create four tart crusts.

Hurrah, our soccer star scored. Hurrah.